ARTICLE III.

TERTIARY VERTEBRATE FAUNAS OF THE NORTH COALINGA REGION OF CALIFORNIA.

A Contribution to the Study of Palæontologic Correlation in the Great Basin and Pacific Coast Provinces.

By JOHN C. MERRIAM.

(Read April 24, 1915.)

CONTENTS.

INTRODUCTION.

One of the most important problems in West-American palæontology and geology concerns the definite determination of time relations between the Pacific Coast marginal marine deposits and the sequence of continental formations of the Great Basin province.   While similar time scales have been used in the regions east and west of the Sierra-Cascade Range, lines distinctly connecting the geological columns of these two regions are rare.   Until recently, remains of the flora have been almost the only means of correlation, but we have known very few satisfactory occurrences of plants appearing in both areas.   Marine invertebrates, found in a remarkably full faunal sequence in the Tertiary marginal marine province, and furnishing the principal materials for correlation between the areas or subdivisions of this province, are absolutely unrepresented in the Tertiary of the Great Basin.   Fresh-water invertebrates are known sparingly in both regions, but the faunal sequence is far from satisfactory, while the characteristic simplicity of these forms reduces the possibilities of correlation.   Land mammals are well represented in the Great Basin province and offer in their stages of evolution a satisfactory basis of correlation.   They have been very rare, however, in the marginal marine areas.

Up to the present time, one of the most significant discoveries throwing light on the problem of correlation between the Pacific Coast and Great Basin provinces is the finding of a considerable number of remains representing land mammals in the marginal marine series of California.   The collection was obtained by a party of advanced students from the University of California, in the North Coalinga region on the western border of the San Joaquin Valley, in December, 1913.   The expedition visited the Coalinga field under the leadership of Dr. Bruce Clark for the purpose of making a study of the sequence of invertebrate faunas, but the discovery of mammal remains seemed so significant that a large share of the work was devoted to examination of the vertebrate horizons.

For the material described in the following paper, the writer is indebted to all of the members of the University of California field party working in the Coalinga region in December, 1913. The first important specimens were obtained in the "Temblor" beds by C. L. Moody and J. M. Douglas. Later all of the members of the party collected at the locality visited by Moody and Douglas. A large part of the material secured at this locality was obtained by Neil C. Cornwall, who systematically excavated for mammalian remains in the soft sandstone and gravel. In the Jacalitos formation, the most important specimens were found by J. H. Ruckman, who worked very carefully over the basal portion of that formation. In the uppermost horizon, good material was obtained by all of the members of the party. The largest portion of this collection was secured by J. O. Nomland, who was making a special study of the Etchegoin. After the return of the December, 1913, expedition, Nomland made a second visit to the region and secured some of the most interesting material representing the latest fauna.

FAUNAL ZONES REPRESENTED.

The vertebrate material obtained in the North Coalinga region represents at least four horizons or faunal zones. Collections were obtained: (1) in the "Temblor" phase of the Monterey series; (2) in the Jacalitos formation; (3) in the lower portion of the Etchegoin formation as mapped in this region by the University ef California party in 1913; (4) a fauna later than that of the lower portion of the Etchegoin section, and obtained in the upper portion of the Etchegoin area.

The lowest horizon is characterized by abundance of horse teeth representing the genus *Merychippus*, and may be known as the Merychippus zone. From the Jacalitos horizon very few remains are known, and there does not seem to be sufficient characteristic material available to warrant final palæontologic designation. The lower Etchegoin horizon is characterized by the presence of *Pliohippus coalingensis*, and may be known as the Pliohippus coalingensis zone. The latest fauna is distinguished by a large specialized horse, with characters of *Pliohippus* and *Equus*, and by remains of a form near *Cervus*. It may be known as Cervus fauna.

The sequence of formations in the region examined is as follows:

Geological Periods.	Local Formations.	Vertebrate Faunas.
Pleistocene	Terraces	
	Tulare	
Pliocene	Etchegoin	Cervus or Odocoileus Pliohippus coalingensis Pliohippus?
	Jacalitos	Neohipparion
Miocene	"Santa Margarita" "Big Blue" Monterey ("Temblor")	{ Merychippus californicus Tetrabelodon? Desmostylus hesperus
Oligocene	Lillis formation[1]	
Eocene	Tejon	
Cretaceous	Chico	

The following generalized section of the region examined by the field party in December, 1913, was prepared for the writer by J. H. Ruckman.

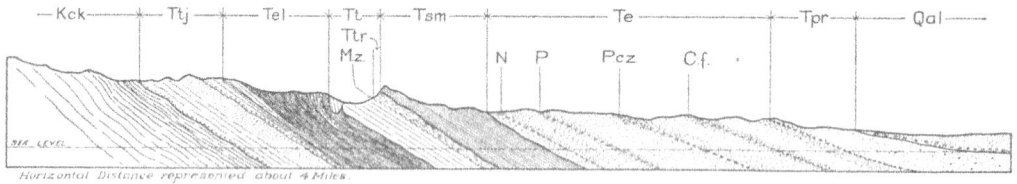

Fig. 1.   Somewhat generalized east and west section across the North Coalinga horizons containing mammalian faunas Section constructed by J. H. Ruckman.   *Kck*, Cretaceous; *Ttj*, Tejon Eocene; *Tel*, Lillis formation, Oligocene; *Tt.*, "Temblor" *Ttr*, "Rainbow beds" or "Big Blue"; *M.z.*, Merychippus zone; *Tsm*, Santa Margarita; *Te*, Jacalitos and Etchegoin; *N*; Neohipparion locality; *P*, Pliohippus? locality; *P.c.z.*, Pliohippus coalingensis zone; *C.f.*, Cervus fauna; *Tpr*, Tulare, formation; *Qal*, Pleistocene.

FAUNA OF MERYCHIPPUS ZONE.

OCCURRENCE.

The vertebrate collection representing the Merychippus fauna was obtained in a zone at the upper limit of the formation known as "Temblor"[2] or "Vaqueros" about twelve miles north of Coalinga.   The best exposures are at locality 2124 in SW. ¼, Sec. 28, T. 18 S., R. 15 E., M. D. B. & M.

[1] From manuscript of J. H. Ruckman on North Coalinga section.

[2] "Temblor" of F. M. Anderson, *Proc. Calif. Acad. Sci.*, 4th ser., vol. 3, p. 18, 1908.   Correlated with the "Vaqueros" of Hamlin by Arnold and Anderson, U. S. Geol. Surv. Bull. 389, p. 87, 1910.

The bones were found mainly in a zone of sandstone and conglomerate two or three feet thick immediately below the beds known as the "Big Blue," which over-lies the characteristic " Temblor " beds.   The zone containing bones consists in part of conglomerate, including pebbles ranging up at least to one inch in diameter. The upper portion of the zone is mainly fine sand.   Fragments of bones and teeth have also been found scattered through the "Temblor" sandstone to a dis-tance of at least one hundred feet below the conglomerate-sandstone zone in which the Merychippus fauna was obtained.

Question has naturally arisen as to the stratigraphic relation of the conglomerate and sandstone of the Merychippus zone to the "Temblor" beds below, and to the "Big Blue" above, as also regarding the relation of the "Big Blue" to the "Temblor."

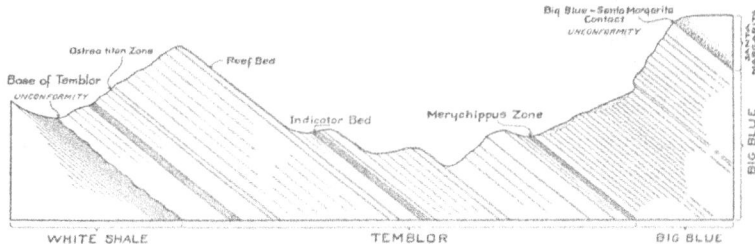

Fig. 2.   Detailed section across the "Temblor" showing location of the Merychippus zone.   Section prepared by C. L. Moody.   Length of section approximately 1225 feet.

The work of the University party in December, 1913, seemed to show a gradual transition from the "Temblor" beds through the Merychippus zone up at least to the base of the "Big Blue."   At some localities there is an irregular contact between the conglomerate and the underlying "Temblor" sandstone, but such relief as was ob-served is possibly not more than should be expected in an estuarine deposit in which coarse sediments were moved at varying rates at different times.   The general con-sensus of opinion of those members of the University party who examined this section was that no important stratigraphic break exists between the characteristic "Tem-blor" portion of the sequence and the top of the Merychippus zone.

The beds north of Coalinga designated as the "Big Blue" were referred to by F. M. Anderson[1] as having the stratigraphic position of the Monterey series.   An-derson stated that " their separation from the Temblor in the fields north of Coalinga is for convenience in logical treatment rather than for emphasis of their stratigraphic prominence."

Arnold and Anderson[2] mapped the " Big Blue " tentatively with the " Santa Mar-

[1] Anderson, F. M., *Proc. Calif. Acad. Sci.*, 4th ser., vol. 3, p. 21, 1908.

[2] Arnold, R., and Anderson, R., U. S. Geol. Survey, Bull. 398, pp. 76, 78, 82, 88, 89; 1910.

garita" but noted that it might really represent a part of the Monterey or a portion of the "Vaqueros" (p. 78). They called attention to its stratigraphic position corresponding to that of the Monterey (p. 76); to the probability that the line at its base may represent a great unconformity (p. 82); and to the overlap of the Tamiosoma zone upon it, indicating that it is a distinct unit (pp. 88 and 89).

In reporting on the Cantua-Panoche region north of Coalinga, after more careful study of the field than was possible in connection with the Coalinga report, Robert Anderson[1] states that in the area north of Coalinga mapped by Arnold and Anderson there is evidence indicating that the "Big Blue" belongs with the "Vaqueros." This evidence includes the presence in the "Big Blue" of marine fossils of early Miocene or "Vaqueros" type. Anderson suggests (p. 64) the possible equivalence of the "Vaqueros" of the Cantua-Panoche region to the lower portion of the Monterey shale in the region nearer the coast.

R. W. Pack, of the United States Geological Survey, who worked carefully over this region with Robert Anderson, states that the fauna of the marine beds below the "Big Blue" occurs also in sandy strata above the typical serpentinous shale, which forms the major part of the "Big Blue." This indicates that the time of deposition of the "Big Blue" belongs to the same biologic period as that of the "Temblor" or "Vaqueros" below, and that there was no great time interval between the deposition of the two.

COMPOSITION OF THE MERYCHIPPUS FAUNA.

The material obtained in the Merychippus zone consists of scattered teeth and skeletal elements. No connected parts of the skeletons, and no teeth so closely associated as to indicate their certain reference to the same individual, have been found. The specimens were all obtained from coarse sands and gravels, presumably representing the wash upon a beach near the mouth of a river. The number of specimens secured exceeds that obtained at any other locality of this nature known to the writer in the Pacific Coast region. The occurrence at this place presumably indicates that when these beds were being deposited the land mammals found here were abundant in this region. It also indicates an unusual combination of circumstances making possible the preservation of these remains.

Associated with the remains of land mammals in the Merychippus zone are several specimens representing marine types. At least three species of sharks appear in this collection. At a level about two feet above the layer in which *Merychippus* teeth were most abundant, Mr. Cornwall obtained a perfectly preserved tooth of the sirenian, *Desmostylus*. As the remains of *Desmostylus* are characteristic fossils of

[1] Anderson, R., U. S. Geol. Surv. Bull. 431-A, p. 65, 1910.

the "Temblor" in this region, question has naturally arisen whether this tooth may not have been derived from the "Temblor" and secondarily deposited in the conglomerate at the base of the "Big Blue." Remains of *Desmostylus* are by no means abundant at any horizon, and it seems improbable that the occurrence of this tooth above the *Merychippus* stratum is the result of secondary deposition. It is probable that this animal was living in the sea in the period in which *Merychippus* inhabited the adjacent land.

The fauna obtained from the Merychippus zone includes the following forms:

> *Merychippus californicus*, n. sp.
> *Prosthennops?*
> *Procamelus?*, sp.
> *Tetrabelodon?*, sp.
> *Desmostylus*, near *hesperus* Marsh.
> *Isurus*, sp.
> *Carcharodon*, sp.

MERYCHIPPUS CALIFORNICUS, N. SP.

Type specimen, M^1, no. 21247, locality 2124. From Merychippus zone between typical "Temblor" and "Big Blue."

Cheek-teeth of the *Merychippus* type, but tending to a more slender form than in *M. isonesus* of the Mascall Middle Miocene.

Remains representing Equidæ from locality 2124 comprise forty or fifty well-preserved upper and lower molars, several incisors, and limb elements including the calcaneum, astragalus, second phalanx, and a metapodial. The upper and lower molars seem all referable to a single species.

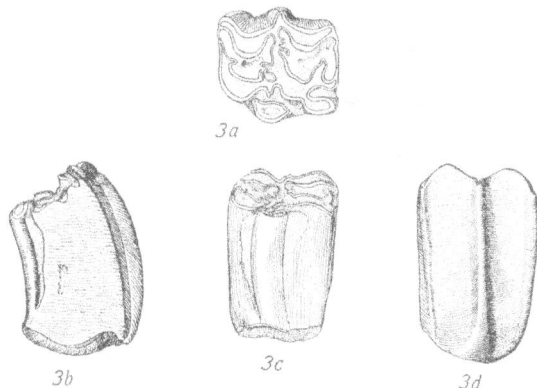

FIGS. 3a–3d. *Merychippus californicus*, n. sp. M^1, no. 21217, natural size, Merychippus zone, North Coalinga region, California. Fig. 3a, occlusal view; Fig. 3b, posterior view; Fig. 3c, inner view; Fig. 3d, outer view.

The characters of the upper molars (Figs. 3a to 8b) are in some respects similar to those of *Merychippus isonesus* of the Mascall Middle Miocene. The protocone is distinct from the protoconule and remains separate almost to the base of the crown.

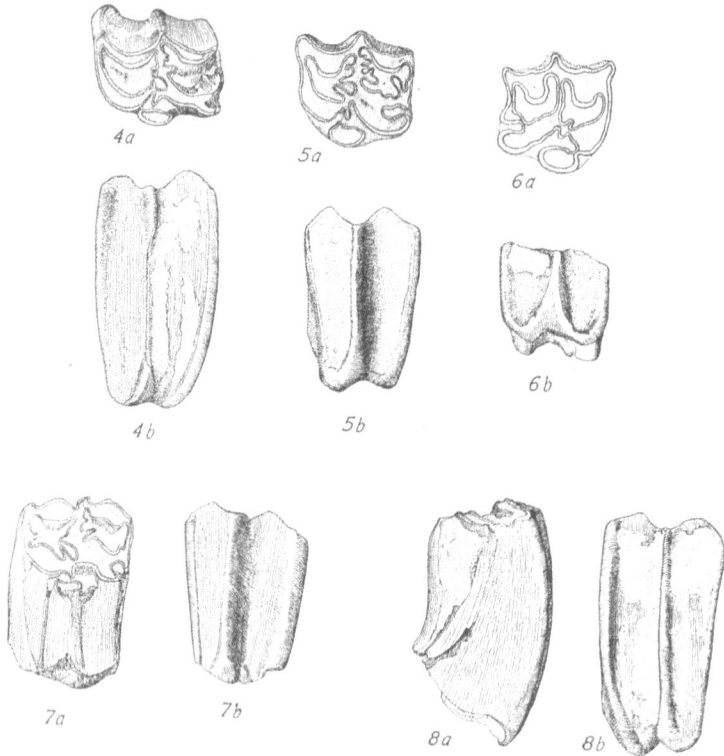

FIGS. 4a–8b. *Merychippus californicus*, n. sp. Upper molars, natural size, Merychippus zone, North Coalinga region, California.

FIGS. 4a and 4b, P^4, no. 21280; Fig. 4a, occlusal view; Fig. 4b, external view.

FIGS. 5a and 5b, M^1, no. 21236; Fig. 5a, occlusal view; Fig. 5b, external view.

FIGS. 6a and 6b, M^1 ?, no. 21241; Fig. 6a, occlusal view; Fig. 6b, external view.

FIGS. 7a and 7b, P^4, no. 21230; Fig. 7a, occlusal and inner view; Fig. 7b, external view.

FIGS. 8a and 8b, M^1, no. 21238; Fig. 8a, anterior view; Fig. 8b, external view.

Only in very old, much worn teeth is there any suggestion of connection of the protocone and protoconule on the crushing surface. In cross-section, the protocone is round with a slight lateral compression. A ridge on the antero-external wall of the protocone reaches outward toward the protoconule, but does not unite with it until an advanced stage of wear is reached. The moderately complex folding of the enamel enclosing the fossettes approximates that of *M. isonesus*. In form and strength of the outer styles the Coalinga form is not appreciably different from typical Mascall specimens. Compared with *M. isonesus*, the upper molars from Coalinga show a tendency to be a little smaller in cross-section, and a little larger in length of crown. The type of upper molar seen here is that of a representative of *Merychippus* in the division of the group leading to *Hipparion*. In this group *M. isonesus* of the Middle Miocene seems to be a less advanced stage and *M. calamarius sumani* of the Barstow Upper Miocene is more advanced.

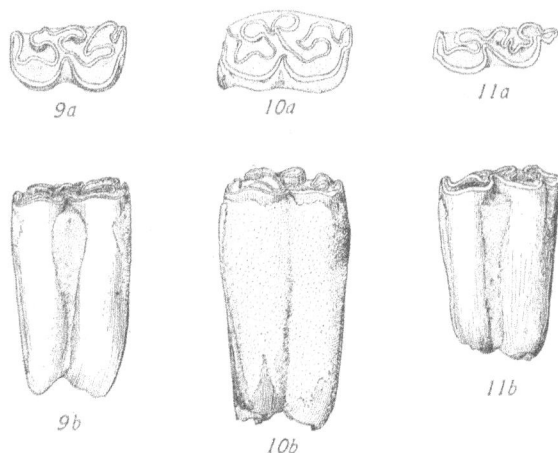

FIGS. 9a–11b.   *Merychippus californicus*, n. sp.   Lower molars natural size, Merychippus zone, North Coalinga region, California.

FIGS. 9a and 9b, P₃, no. 21244; Fig. 9a, occlusal view; Fig. 9b, external view.

FIGS. 10a and 10b, P₄, no. 21508; Fig. 10a, occlusal view; Fig. 10b, external view.

FIGS. 11a and 11b, M₂, no. 21249; Fig. 11a, occlusal view; Fig. 11b, external view.

The lower cheek-teeth (Figs. 9a to 11b) from Coalinga closely resemble *M. isonesus* in form and dimensions. The Coalinga specimens are possibly a little longer, but the difference is small. The metaconid-metastylid column is moderately flattened, and an antero-external ridge may be present on the protoconid.

Merychippus seversus, a form apparently widely distributed in the Mascall Miocene, is clearly separable from the Coalinga species by its much smaller size.

Merychippus calamarius, a type abundantly represented in the Barstow Upper Miocene of the Mohave Desert, is larger and more advanced. One form, *M. calamarius sumani*, occurring rather rarely in the Barstow fauna approaches the combination of characters seen in the Coalinga specimens, but is a more advanced type, and is specifically separable.

Two incisor teeth (Figs. 12a to 13b) found with the Coalinga *Merychippus* fauna show deep invaginations of the enamel, and a thick cement deposit in the pits.

FIGS. 12a–13b. *Merychippus californicus*, n. sp. Incisors, natural size. Merychippus zone, North Coalinga region, California. Figs. 12a and 12b, no. 21381; Fig. 12a, anterior view; Fig. 12b, occlusal view. Figs. 13a and 13b, no. 21374; Fig. 13a, anterior view; Fig. 13b, occlusal view.

Of the protohippine limb elements (Figs. 14a to 14d) from locality 2124, a single calcaneum is slightly larger than those referred to *Merychippus* from the Mascall and Virgin Valley. It is of approximately the size of this element in *Merychippus calamarius* from the Barstow beds. An astragalus has dimensions approximately the same as the average *Merychippus* specimens from the Mascall and Virgin Valley. Another much worn specimen is considerably smaller and corresponds in size to the smallest available astragalus from the Mascall stage. The smaller astragalus is not as large as the smallest one known from the Barstow beds. The larger specimen slightly exceeds the dimensions of the smallest from the Barstow, but is considerably below the average size.

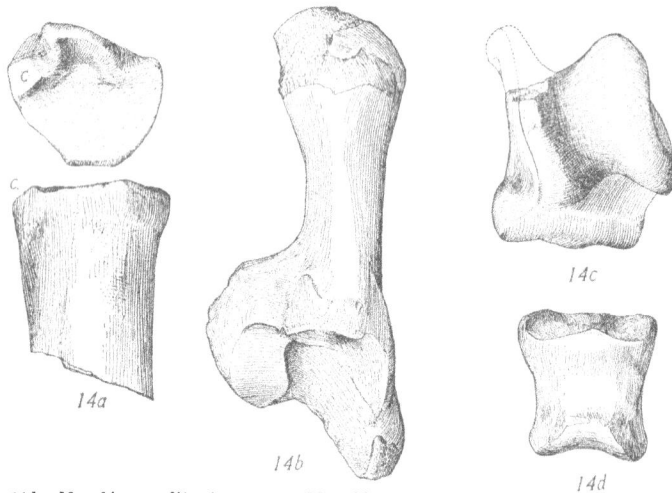

FIGS. 14a–14d. *Merychippus californicus*, n. sp. Merychippus zone, North Coalinga region, California. All figures natural size. Fig. 14a. Proximal end of metatarsal III, no. 21509, anterior and proximal views; c, cuboid facet. Fig. 14b. Calcaneum, no. 21367, superior view. Fig. 14c. Astragalus, no. 21368, superior view. Fig. 14d. Second phalanx, no. 21366a, superior view.

A metatarsal III (Fig. 14a), no. 21509, from the Merychippus zone is of approximately the same size and form as the corresponding element in *Merychippus* specimens from the Mascall beds of Oregon.   The Merychippus zone specimen is perhaps a little thicker anteroposteriorly.   The facet for the cuboid is at least as large in no. 21509 as in the Mascall specimens, and is if anything a little nearer the plane of the ectocuneiform facet.   The form of the proximal end of metatarsal III in no. 21509 can be matched almost exactly in *Merychippus* specimens from the Barstow fauna, but some of the Mohave specimens are larger and have a larger cuboid facet.   The greatest transverse diameter of the proximal end is twenty-six millimeters,

The phalanges correspond in dimensions to those of the Mascall and Virgin Valley horses and are smaller than in the average *Merychippus* forms of the Barstow beds.

MEASUREMENTS OF CHEEK-TEETH.[1]

All measurements of anteroposterior and transverse diameter were taken about ten millimeters above the base of the crown.

Height of crown as given is the greatest height measured along the mesostyle of upper teeth or along the hypoconid of lower teeth.

	No. 21250 P^2	No. 21326 P^3	No. 21240 P^4	No. 21239 P^4	No. 21238 M^1	No. 21236 M^1	No. 21237 M^1	No. 21246 M^2
Anteroposterior diameter..	21 mm.	20.3	17.8	19	17	18.3	19	15.2 with cement
Transverse diameter......	18	a21	a20.5	20.8	21	20.2	19.9	17.7 with cement
Height of crown.........	22.5 worn	35.5	36	27 worn	37.4	25.9 worn	28.5 worn	29.5

	No. 21325 P_2	No. 21245 P_3	No. 21508 P_4	No. 21249 M_2	No. 21324 M_3	No. 21329 M_3
Anteroposterior diameter........	20 mm.	19.3	21.2	16.7	20.5	21.5
Transverse diameter.............	9.9	10.2	10.4	9	7.5	a8
Height of crown.................	16.5 worn	33.7	38.8	28.5	21.5 worn	33.4

PROSTHENNOPS?, SP.

A single lower molar (no. 21511, Figs. 15a and 15b) from the Merychippus zone represents a peccary near *Prosthennops*.   The tubercles are low and less acute than in *Bothriolabis* of the John Day.

[1] For method of measurement see Univ. Calif. Publ. Bull. Dept. Geol., vol. 7, p. 409, 1913.

MEASUREMENTS OF No. 21511.

mm.

? $M_{\overline{2}}$, greatest anteroposterior diameter... 14.5

? $M_{\overline{2}}$, greatest transverse diameter... 11.4

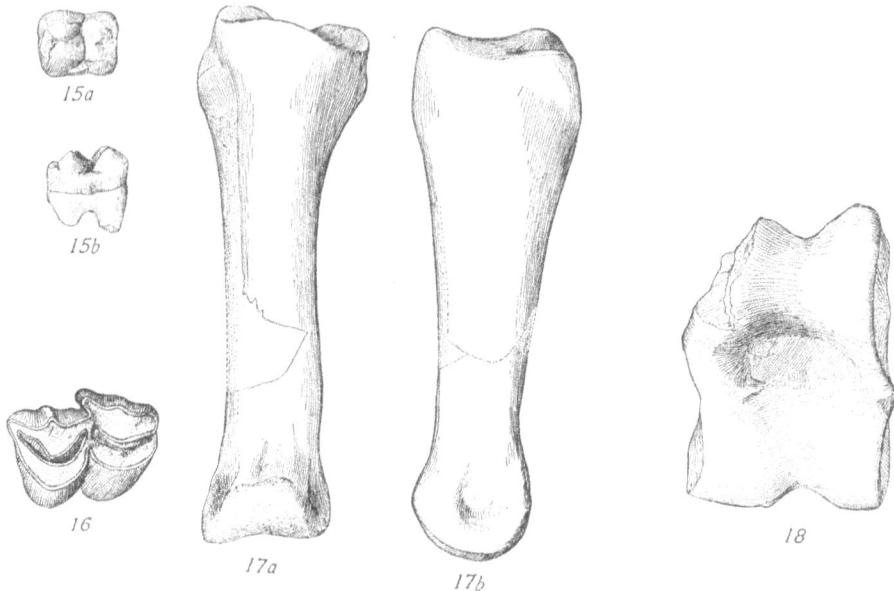

FIGS. 15a and 15b. *Prosthennops* ?, sp. M_2 ?, no. 21511, natural size. Merychippus zone, North Coalinga region, California.

FIGS. 16–18. *Procamelus* ?, sp. Merychippus zone, North Coalinga region, California. All figures natural size, Fig. 16, M^2, no. 21510, occlusal view. Figs. 17a and 17b, proximal phalanx, no. 21343; Fig. 17a, superior view; Fig. 17b, medial view. Fig. 18. Astragalus, no. 21344, superior view.

PROCAMELUS ?, sp.

The remains referred tentatively to *Procamelus* (Figs. 16 to 18) consist of several somewhat worn astragali, a portion of a calcaneum, a fragment of the distal end of a metapodial, a proximal phalanx and an upper molar. These specimens are near the size and dimensions of *Procamelus* forms from the late Miocene. The proximal phalanx (no. 21343) is much larger than that of the Recent llama. It is like that of *Procamelus*, but is larger than in some forms in the later Miocene. The astragali are a little larger than in the Recent llama, and smaller than in many specimens of *Procamelus*. An astragalus from locality no. 2028, in the Cedar Mountain region of Nevada, is similar in form, but a trifle larger. The astragali (no. 21344 especially) are almost identical in form with certain small cameloid astragali from the Upper Miocene of the Mohave Desert which are tentatively referred to *Procamelus*.

A single upper molar tooth, no. 21510, constitutes the only representation of the dentition.

MEASUREMENTS OF UPPER MOLAR No. 21510.

	mm.
? M^2, greatest anteroposterior diameter	26.2
? M^2, greatest transverse diameter	15.9

FIGS. 19a–20b.   *Tetrabelodon* ?, sp.   Cheek-teeth, × ½.   Merychippus zone, North Coalinga region, California. Figs. 19a and 19b, occlusal and end view of fragment of cheek-tooth no. 21366.   Fig. 20a, occlusal view of milk molar no. 21377; Fig. 20b, lateral view of specimen 21377.

TETRABELODON?, sp.

A considerable number of fragments represent a proboscidean near *Tetrabelodon* (Figs. 19a to 20b).   The only complete tooth is a milk molar.   Though certain generic determination does not seem possible, the absence of all proboscidean remains from beds older than Middle Miocene in America makes these specimens of importance in determining the epoch that this fauna could be presumed to represent.

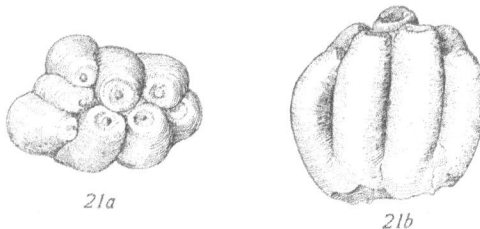

FIGS. 21a and 21b.   *Demostylus*, near *hesperus* Marsh.   M¹, no. 21375 × ½.   Merychippus zone, North Coalinga region, California.   Fig. 21a, occlusal view; Fig. 21b, lateral view.

DESMOSTYLUS, NEAR HESPERUS MARSH.

Two well-preserved cheek-teeth, an upper molar (Figs. 21a and 21b), and a lower molar, represent *Desmostylus*, the peculiar sirenian now known to have ranged widely around the North Pacific in Middle Tertiary time.   In an earlier article[1] the writer has assembled all available information relating to *Desmostylus*, with the result that the best known occurrences seem to fall in the zone of the "Temblor" or "Vaqueros" of California.   Other occurrences are apparently near that horizon. If the vertical range of this form is not unusually long, the wide geographic range

[1] Merriam, J. C., Univ. Calif. Publ. Bull. Dept. Geol., vol. 6, pp. 403–412, 1911.

through marine deposits of the North Pacific gives us an almost unprecedented opportunity for correlation, as the horses with which *Desmostylus* occurs in these beds have a wide geographic range over the land.

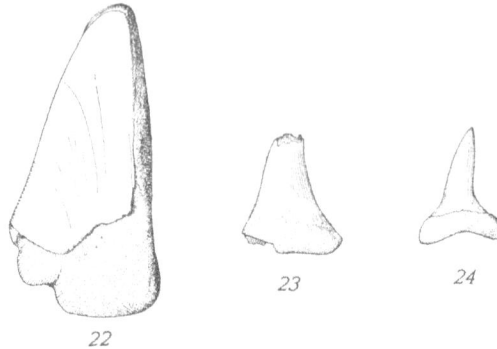

FIGS. 22–24. Selachian teeth from the Merychippus zone, North Coalinga region, California. All figures natural size.
FIG. 22. *Carcharodon*, sp. no. 21563.
FIG. 23. *Isurus*, sp. no. 21565.
FIG. 24. *Isurus?*, sp. no. 21564.

SELACHIAN TEETH.

Three types of shark teeth from the beds at locality 2124 represent forms corresponding approximately to species occurring in strata referred to the "Temblor" horizon of middle and southern California. One specimen (Fig. 22) is a fragment of a *Carcharodon* tooth from which the denticles have been removed by wear in shifting about in the sand before final burial. A second worn specimen (Fig. 23) is evidently a form of *Isurus;* a third (Fig. 24) is a worn *Lamna* or an *Isurus*.

STRATIGRAPHIC AND PALÆONTOLOGIC CORRELATION OF THE MERYCHIPPUS ZONE.

The appearance of a *Merychippus* fauna in the essentially marine section of the California area is of significance with respect to the broad problem of time relations of the West-American Tertiary. In considering the meaning of this occurrence, it is desirable: *first*, to determine as certainly as possible the position of the Merychippus zone in the marine section of the California area and to ascertain the basis for determination of age of this particular member of the California marine series; *second*, to determine the relation of the Merychippus fauna to those in the mammalian sequence of America, and the time relations of the American mammalian series to the succession in typical sections of the Old World.

Position of Merychippus Zone in the California Marine Series.—As has been shown in the general discussion of occurrence of the Merychippus fauna, the evidence indicates that this zone is included within a portion of a series designated as "Temblor" by F. M. Anderson and referred to as "Vaqueros" in all recent accounts of this region published by members of the United States Geological Survey. The strata referred to the "Temblor" or "Vaqueros" of the North Coalinga field are evidently a part of a wide-spread series of deposits generally recognized as representing one great period of sedimentation and designated as the Monterey series.[1]

In the earliest attempt at separation of faunal zones in the section between the Tejon Eocene and the San Pablo Miocene the writer[2] contrasted the fauna of the lowest portion of the beds referred to the Monterey series in the Contra Costa Hills with that of higher members of that section, exclusive of the contrast between sandstone and shale members due to varying conditions of deposition. The faunal zone of the lowest beds above the Tejon Eocene of the Contra Costa Hills was designated the Agasoma gravidum zone, and was considered as Lower Miocene. This fauna was compared with that of beds presumed to be of Lower Miocene age in the southern part of the state. Two faunal assemblages from the southern area were referred to the Lower Miocene. One assemblage, characterized by the presence of *Turritella ocoyana* and a number of *Agasoma* species, was referred to as the Turritella ocoyana zone. The other fauna, characterized by *Turritella inezana* and an *Agasoma*, was distinguished as the Turritella (hoffmani) inezana[3] zone. It was suggested that these zones did not represent the same horizon, but that they were not widely separated. The ocoyana zone was shown to have a fauna of more recent aspect than that of the inezana zone. The *Agasoma* forms of the ocoyana zone were described as in some respects intermediate between *Agasoma gravidum* of the Contra Costa Hills, and *Agasoma sinuatum*, a species from the upper portion of the Contra Costa Hills section.

Following shortly after the faunal separation of the Agasoma gravidum zone, Hamlin[4] described as the "Vaqueros formation," a section containing the fauna of the Turritella inezana zone. Later, from the southern part of the state, F. M. Anderson[5] described as the "Temblor formation" an important series of strata representing in part at least the Turritella ocoyana zone.

F. M. Anderson and Arnold and Anderson have considered the "Vaqueros"

[1] Lawson, A. C., Univ. Calif. Publ. Bull. Dept. Geol., vol. 1, pp. 1–59, 1893. See also Louderback, G. D., Univ. Calif. Publ. Bull. Dept. Geol., vol. 7, pp. 177–241, 1913.

[2] Merriam, J. C., Univ. Calif. Publ. Bull. Dept. Geol., vol. 3, pp. 377–381, March, 1904.

[3] *Turritella hoffmani* Gabb has been shown by Arnold to be identical with the previously described *T. inezana* of Conrad.

[4] Hamlin, H., U. S. Geol. Surv. Water Supply Paper No. 89, distributed June 15, 1904. Hamlin's use of name preceded by H. W. Fairbanks, U. S. Geol. Surv. Folio 101, distributed June 10, 1904.

[5] Anderson, F. M., *Proc. Calif. Acad. Sci.*, 3d ser., Geol., vol. 2, p. 170, 1905.

of Hamlin and the "Temblor" of F. M. Anderson as practically representing the same stratigraphic and faunal unit.   Arnold and Anderson, by virtue of their decision as to the stratigraphic identity of the deposits containing these two faunas, have used the name "Vaqueros"[1] as having precedence in use over "Temblor."   F. M. Anderson,[2] also holding that the two zones represent the same stratigraphic unit, considers the description of the "Vaqueros" presented by Hamlin as insufficient.   In reality the description by Fairbanks, based on a typical section, precedes that of Hamlin and establishes the name.

The most recent investigations of the stratigraphic, geographic, and faunal relations of the Turritella inezana and Turritella ocoyana zones seem to the writer still to suggest that these zones may represent recognizable stages in the evolution of the Tertiary marine faunas of the California area, and that there is a possibility of stratigraphic separation of the two.   Under the circumstances, the writer is inclined to continue his use of the faunal designations of these horizons suggested in 1904. If the faunal separation is continued, and if the stratigraphic relations are not clear, it seems desirable to use for geologic reference to the strata containing these faunas terms which suggest their geographic occurrence.   The writer is therefore inclined for the present to use the name "Vaqueros" for the series of strata containing the *Turritella inezana* fauna.   The name ".Temblor" is tentatively used for those sections in the southern part of the state containing the *Turritella ocoyana* fauna, on the presumption that this part of the series may be stratigraphically distinct from the "Vaqueros."   It is probable that the "Temblor" as thus designated is really a part of the Monterey series as now recognized.

In a recent discussion of the geologic range of Miocene faunas of California, Professor James Perrin Smith[3] has taken the view that the "Temblor" fauna was synchronous with that of the Monterey, and that the Turritella inezana zone represents a horizon below the typical Monterey.   The name "Vaqueros" is used by Professor Smith for the beds containing the *Turritella inezana* or "Vaqueros fauna."

The available evidence indicates that the Merychippus zone occupies a position within the "Temblor," and presumably some distance below the upper limit of the Monterey series.   If the beds of this zone were considered unconformable upon the "Temblor," they would presumably correspond in time equivalence to some portion of the upper Monterey beds as represented in middle California. The "Big Blue" resting above the Merychippus zone is separated by marked unconformity from the "Santa Margarita" above it.   The "Santa Margarita" of the

[1] Arnold, R., and Anderson, R., U. S. Geol. Survey Bull. 398, p. 87, 1910.

[2] Anderson, F. M., *Proc. Calif. Acad. Sci.*, 4th ser., vol. 3, pp. 38 and 39, 1908.

[3] Smith, J. P., *Proc. Calif. Acad. Sci.*, 4th ser., vol. 3, pp. 164, 165 and 169, 1912.

Coalinga region, according to Dr. Bruce Clark,[1] contains a fauna closely similar to that of the middle San Pablo.   The unconformity between the "Santa Margarita" and "Big Blue" apparently marks a long period, which probably represents a time sufficiently long to equal the period of deposition of the earlier portion of the San Pablo.

Age of the Monterey.—During several decades past the Monterey series has been considered as Miocene and strata containing a faunal assemblage of the stage seen in the "Temblor" of the Coalinga region have been generally recognized as Lower Miocene.   In actual practice the determinations of age of strata referred to the Miocene have never been based on very extensive or exact comparisons.

In describing a collection of fossils from Astoria in 1849 Conrad[2] expressed his view as to age determination of fossiliferous deposits as follows:

". . . the forms are decidedly approximate to those of the Miocene period which occur in Great Britain and the United States. *Nucula divaricata*, for instance, closely resembles *N. Cobboldiæ* (Sowerby) of the English Miocene, and *Lucina acutilineata* can scarcely be distinguished from *L. contracta* (Say), a recent species of the Atlantic Coast and fossil in the Miocene beds of Virginia. *Natica heros*, a shell of similar range, is quite as nearly related to the *N. saxea*. A similar number of species might be obtained from some of the Miocene localities of Maryland or Virginia and yet no recent species be observed among them. In the Eocene, and also in the Miocene strata, there are peculiar forms which obtain in Europe and America, and although the species differ, yet they are so nearly allied that this character alone, independent of the percentage of extinct forms, is quite a safe guide to the relative ages of remote fossiliferous rocks. On this foundation, I speak with confidence, when I assign the fossils of the Columbia River to the era of the Miocene."

In one of the first references to the Miocene of California, Conrad,[3] in speaking of beds at Santa Barbara, called Miocene, expresses the following view as to the age of the strata:

"The *Mercenaria* and *Pecten* are closely related to species of the Virginia Miocene, and indeed there is an extraordinary analogy in all of the above mentioned shells to species of the Atlantic Miocene deposits. . . ."

Discussing the age of fossil-bearing deposits in California in 1857 Conrad[4] makes the following statement:

"Like the Miocene of Virginia, the Estrella group is characterized by large and even comparatively gigantic species of Pectinidæ, so unlike any living on the coast of California or the Atlantic

[1] Clark, B., Unpublished manuscript on the Fauna of the San Pablo Miocene.

[2] Conrad, T. A., U. S. Exploring Expedition, 1849, p. 659 (see original reference, *Am. Jour. Sc.*).

[3] Conrad, T. A., *Proc. Acad. Nat. Sc. Philad.*, vol. 7, p. 441, 1834.

[4] Conrad, T. A., Pacific Railroad Survey, vol. 7, p. 189, 1857.

states. It would seem that this family had then reached their maximum of development and the genus *Pallium* was first introduced, and of far larger size than any which now exists. It is worthy of remark that the generic character is developed on a far grander scale than appears in subsequent epochs, the prominent teeth and thick hinge reminding us of the genus *Spondylus*.

"Every new collection of Miocene fossils shows more clearly the connection between some of the Tertiary strata of California and those of Virginia. The species in the present collection are far more interesting than in others of the same formation on the Pacific slope which I have yet seen. It does not appear that this group of fossils has any living representative in the present fauna of the Pacific Coast, but several of them approximate to extinct Virginia species; and I am not sure that the large *Pecten magnolia*, herein described, is not identical with the Virginia species *P. jeffersonius*.

"I think it might safely be assumed that the San Rafael Hills, Santa Inez Mountains and Estrella Valley contain strata which are parallel to Miocene sands and clays of the James and York rivers in Virginia."

.   Later writers have generally followed the lead of Conrad in fixing the age of middle Tertiary strata, including the Monterey. In an early descripion of the type locality of the Monterey, Blake[1] simply refers to it as Tertiary. In the description of these beds by Lawson[2] the statement is made that characteristic Miocene fossils have been found in the Monterey series at various parts of the coast by former observers, and in particular at the town of Monterey. A list of species obtained from a locality near Carmelo Bay, and determined by W. H. Dall, is published in Professor Lawson's paper.

In recent years, the results of work on the Tertiary of California have brought to light previously unknown thicknesses of strata, new and important unconformities indicating large gaps in the record, and new faunal zones intercalated between those previously known. The addition of these factors, which lengthen the geologic record, has placed in the division recognized as Miocene time an unexpectedly long series of events, and has very naturally raised a question whether the lowest beds referred to the Miocene do not represent Oligocene. In a discussion of the geologic range of *Desmostylus*,[3] a sea-cow, in 1911, the writer described this form as most common in beds called Lower Miocene of the southern part of the state, and suggested that the beds marking its downward limit of geologic range might correspond to Oligocene.

Recently Arnold[4] and Hannibal, in a classification of the Tertiary formations of the Pacific Coast region, have stated that the " Monterey . . . as far as our present knowledge goes, might be placed equally well in the latest Oligocene or the earliest Miocene on the basis of the general faunal facies."

[1] Blake, W. P., Pacific R. R. Report, vol. 5, chapter 13, 1856.

[2] Lawson, A. C., Univ. Calif. Publ. Bull. Dept. Geol., vol. 1, p. 27, 1893.

[3] Merriam, J. C., Univ. Calif. Publ. Bull. Dept. Geol., vol. 6, p. 407, 1911.

[4] Arnold and Hannibal, H., *Proc. Amer. Phil. Soc.*, vol. 52, p. 575, 1913.

Though practically all determinations of the age of marine Tertiary formations in California have been based upon palæontologic correlation, the true basis of comparison of these horizons has never been clearly expressed. The determinations were made either by very general correlation with determined faunas of eastern United States and Europe, or by use of the Lyellian percentage method.

The comparisons of West Coast faunas of post-Eocene age with those of the Atlantic area, upon which the original age determinations of Conrad and Gabb were based, were made at a time when the faunas of the Atlantic area were not as well known as at present, and the study of the Pacific Coast faunas was hardly begun. The results of a thorough comparison with the Atlantic faunas are not available to us as the basis for a fully satisfactory judgment from this point of view.

In connection with a recent study of the marine Tertiary of the Washington-Oregon area, Arnold and Hannibal[1] express the opinion that "a direct correlation between the Pacific Coast marine Tertiary and the deposits of Europe and bordering the Gulf of Mexico is impossible owing to almost total absence of identical species except in the Eocene." In consideration of possible correlation of strata on the Pacific Coast region with those of the Oligocene of Europe, attention is called to the fact that there do not appear to be any marine forms that will serve as a basis for direct comparison. The suggestion is, however, made that an Oligocene facies is indicated by absence of all Recent molluscan species from the older strata considered as post-Eocene, and the gradual addition of a small percentage of Recent species in the Monterey. The presence of certain Eocene-Oligocene genera as *Crassitellites*, *Aturia*, etc., also suggests Oligocene age.

A number of age determinations of Pacific Coast Tertiary formations have been based upon use of the percentage method, and while fairly consistent results were secured, it must be borne in mind that determinations obtained by this means may be subject to modification by several factors. It is, for example, evident that since the time of Lyell views as to the limits of species have greatly changed, and the percentages used for the species of Lyell's day would not apply with the same result upon the same faunas with present-day limitations of species. Species as now defined are much more closely restricted in range of characters. This limitation of the characters in specific groups has naturally restricted also the geographic and geologic range. Studies in the general accuracy of percentage method age-determinations recently made by Bruce Martin at the University of California have shown surprisingly large errors from various sources in ordinary computations, unless a very full representation of the fauna is available.

[1] Arnold, R., and Hannibal, H., *Proc. Amer. Philos. Soc.*, vol. 52, p. 574, November–December, 1913.

Aside from purely palæontologic determinations of the age of Tertiary formations in the California area, we have practically nothing as a basis for definition of historic stage in comparison with the later portion of geologic history in other parts of the world. Though every one admits the exceeding desirability of some method of comparison connected solely with a study of crustal movements, scarcely any attempt has been made to apply such a means of determination for stratigraphic units of this region older than the close of the Pliocene. Presuming that we have in the Pleistocene a satisfactory medium of comparison through crustal movement, without reference to climatic and faunal characters which are commonly the means of correlation employed, there seems no satisfactory method of comparison for systems older than Pleistocene other than by approximate matching of unconformities.

The California Tertiary geologic sequence is notoriously complicated. Crustal movements have been frequent; unconformities are common; and it is difficult to see how any one working in California geology up to the present time could have selected the particular unconformities or determined the movements needed to give us the key to comparison of our Tertiary history with that of eastern United States or of Europe. At any rate, this has not been done with any assurance of accuracy. At best there could be nothing more than a rough approximation based on the assumption that some of the diastrophic events here have a definite relation to those that have occurred elsewhere.

A general consideration of the present situation with reference to determination of age of the members of the California Tertiary sequence makes it evident that our present determinations are not based on thoroughly satisfactory studies in any of the several possible fields of endeavor, and that much work may profitably be given to an examination of this problem.

Position of the Merychippus Fauna in the Mammalian Sequence of America.— In terms of the vertebrate series of western North America, the fauna of the Merychippus zone in the North Coalinga region is clearly later than Lower Miocene, and not later than Upper Miocene. Proboscidean remains are not certainly known in America at a horizon lower than Middle Miocene. The wide-spread genus *Merychippus* is not known earlier than Middle Miocene, and the characteristic forms of that genus in the Middle Miocene of America are not more advanced than the species of the Merychippus zone in the Coalinga region. The number of horse remains found gives ample opportunity to determine the stage of evolution of the cheek-tooth dentition, and it seems scarcely possible that there could be any misunderstanding as to the stage in the history of this group represented here. The available material representing camels does not suggest an age determination different from that based on the representatives of the horse and mastodon groups.

FIG. 25. Sketch map of Pacific Coast and Basin regions showing geographic divisions based on distribution of Tertiary faunas.

Whatever may be the relation of the divisions of the American continental Miocene fauna or formation sequence to the sequence of the Old World based upon the succession of mammalian forms, it seems clear that the Merychippus zone of the Coalinga region is near the stage of the Great Basin Middle Miocene fauna represented in the Mascall of eastern Oregon and the Virgin Valley of the northern Nevada region. Through the medium of comparison with the Great Basin sequence, there is justification for correlating the stage of the Merychippus zone approximately with that of the Pawnee Creek beds of northeastern Colorado and other occurrences of beds referred to Middle or late Miocene in Western North America.

The only means of escape from the conclusions stated above as to relative age of the Coalinga *Merychippus* fauna would presumably be reached through consideration of the possibility that the fauna of the California Merychippus zone originated west of the Wasatch, or that it first appeared in California or possibly in the Great Basin and remained in the extreme west for some time before migrations populated the Great Plains area.

The presence of a form of the mastodon type in the Coalinga Merychippus zone may suggest the possibility that this fauna represents the earliest appearance in America of Old World immigrants not known until considerably later in the region to the east. While it is possible that Asiatic forms colonizing America followed the west coast of the continent south to California, it is unsafe to consider this as necessarily the most inviting path for migration. Other routes leading into the Great Plains region may have been easily accessible. Aside from the California specimens under consideration here, the oldest American representatives of the mastodon group are found in formations situated to the east of the eastern boundary of the Great Basin.

The *Merychippus* group evidently originated in America. The Equidæ were largely represented here in late Oligocene time, and were unrepresented at that stage in the Old World so far as known. Unfortunately for consideration of the possibility of western origin of *Merychippus* types, we have, as yet, no representation of the mammalian fauna of the Great Basin province between the close of the John Day Oligocene and the beginning of the Middle Miocene. As nearly as can be judged, this interval represents a long period. During the earlier portion of this time the John Day beds were subjected to erosion. In the latter part of the interval the Columbia Lava accumulated in the northern portion of the province, and great rhyolite outbursts seem to have occurred in the Middle Basin area. It seems that the Great Basin province may have been a region with somewhat varied relief in which no extensive accumulation of deposits occurred during the first part of the

Lower Miocene, and was widely covered with lava flows in the second part of this epoch.

As *Merychippus* is an animal with tooth and foot structure better adapted to grazing and to plains-living than the known American Lower Miocene horses, it possibly developed in the Great Basin on wide stretches of plains in early Lower Miocene, or on the great lava plains of late Lower Miocene.   So far as environmental conditions have influence in the evolution of specializing types, one might expect to find such a form as *Merychippus* arising in response to the stimulus of environment on the considerably greater areas of the Great Plains country farther to the east, rather than in the Great Basin.

The further possibility is available that *Merychippus* originated in Lower Miocene time on the extreme western border of the continent, in California.

According to the view that the greatest number of new progressive forms will arise on the largest continental masses, one almost unavoidably considers the middle of the continent as the locus of evolutionary or creative activity.   This is true mainly in the sense in which the center of gravity of any mass is the controlling point.   It may be that temporarily isolated marginal areas bordering large land masses furnish at least as satisfactory points of origin or nurseries of new forms as could be provided by any set of conditions on other portions of a large land mass, and possibly the California region has been favorably conditioned for inducing the development of new types of mammals.

The California area, as it appears today, offers on the plains of the Great Valley an unusual combination of conditions which might well be favorable to production of a new type of horse.   In the time of deposition of the Merychippus zone of Coalinga, a large part of the Coast Range region and the southern end of the Great Valley were under water, and the degree of isolation. conditioned by the Sierra boundary was probably less marked than at the present day.   For a considerable time preceding the deposition of the Merychippus zone, conditions were not materially different from those during the deposition of the beds at that horizon.   Somewhat earlier the land area was larger.   The conditions do not, however, seem to have been as favorable for the development of the *Merychippus* group in this area as they probably were in many other regions.   At least, there is no special warrant for supporting the hypothesis that California was the point of origin of this group.

On the whole, the evidence indicates that the fauna of the Coalinga Merychippus zone is at least as advanced as that of the Mascall and Virgin Valley formations of the Great Basin; and that there is no reason for assuming that California was the place of origin of this group.   It also appears that the Mascall and Virgin Valley faunas are not younger than the stage of evolution of assemblages considered to represent the Middle Miocene of the areas to the east, and that there was probably close faunal connection with the region to the east.   The Middle Miocene fauna of the Great Basin evidently represents approximately the same period as that of other areas farther to the east.   There is no reason for believing that the Great Basin fauna in general or the *Merychippus* group in particular necessarily originated in this region.   A portion of the fauna derived from Asiatic immigration may have appeared here before it reached the Great Plains area, but the chances are against its having been largely represented here before it was known elsewhere on the continent.   The possible Californian or Great Basin origin of Miocene groups should not be overlooked in further consideration of the Pacific Coast faunas, but it is improbable that the geographic provinces were so sharply separated in early Miocene time that a group of the *Merychippus* type could become widely spread in the Great Basin and Pacific Coast provinces and remain unrepresented east of these regions.

Relation of American Miocene Mammal Horizons to Those of the Old World.—In correlating marine Middle Tertiary faunas of western North America with those of Europe, it is difficult to make direct comparisons, as the North Pacific Ocean was connected with the marine area bordering Europe only by circuitous routes, or by paths leading through zones of widely varying temperatures.   For this reason, one might expect to find relatively little correspondence, excepting in similarity of stage of evolution of forms surviving from faunas common to both regions at some earlier period.   There is little known in the comparative history of marine life suggesting migration of faunas between these areas in middle Cenozoic time to such an extent as to leave a distinct impression.

In a comparison of the history of the mammalian fauna of North America with that of the Old World, one might expect to find that the grouping of continents around the North Pole has permitted moderate crustal movements to bring about occasional union of those land masses, and that this connection of the areas would invite faunal mingling.

In the critical regions where connection and separation of the continental areas might be expected, especially the Alaskan-Siberian region, there is evidence of varying position of land and water, suggesting alternations of union and separation of the land areas.   Whatever inhibitive influence might be exerted by an arctic climate, such as that of the present day, was largely eliminated throughout the greater part of Tertiary time, as the earth climate was somewhat warmer than at present, and the polar regions did not present an insuperable temperature barrier.

In the Tertiary history of mammalian faunas in the Northern Hemisphere, we note certain epochs marked by similarity of life over nearly the whole region.   These epochs alternate with times of development of provincial faunas.   Without calling in the suggestion of independent origin of similar types in different regions too frequently to satisfy the law of probability, it is difficult to account for these recurrences of similar faunas unless we accept the view that they have been induced by crustal movements uniting and separating land areas, and also closely related to or concurrent with considerable climatic variation.   The opportunity for intermigration of faunas on the land areas concerned appears on the whole better than in the case of marine life in the bordering seas, as the path of migration was shorter and involved less range of latitude.   Other factors being approximately even, in a study of the Tertiary of the Northern Hemisphere, one might expect more satisfactory results in correlation based on comparative studies in the history of land mammals than in comparison of marine invertebrates.

Through the combined work of many investigators, and especially by the comparatively recent assembling of all results in this field by Osborn, Matthew, Scott, Schlosser, and others, it is evident that the portion of Tertiary time, including Oligocene, Miocene, and Pliocene, is actually marked by important intermigrations of land mammals between America and Eurasia.   Within this time, the early Oligocene is characterized by appearance of similar forms in North America and in the Old World.   In late Oligocene, intermigration seems less marked.   In Middle Miocene the movement is particularly noticeable again, especially in the first certain appearance in America of the proboscideans, the teleocerine rhinoceroses, and probably several carnivore groups.   In later Miocene and Pliocene time, the number of common forms increases considerably.

What is known as the Middle Miocene mammalian fauna of America shows some resemblance to that of the Lower Miocene of Europe, and question has been raised as to the time relations of the lower and middle divisions.   In many respects there seems, however, closer correspondence of the American Middle Miocene, as represented by the Mascall and Pawnee Creek faunas, with the Middle Miocene of Europe than there is between these American faunas and that of any European Lower Miocene division.   There is general agreement in the succeeding stages of the two areas, which indicates that the mammalian sequences of North America and Eurasia are brought into approximately the same relation throughout the later Tertiary series.   If any difference exists, or if further adjustment occurs in the scale, it is doubtful whether this modification will do more than to bring to the same level the upper portion of the Lower Miocene of Europe and the lower portion of the Middle Miocene of America.

SIGNIFICANCE OF THE MERYCHIPPUS ZONE OF COALINGA WITH RELATION TO THE PROBLEM OF CONSTRUCTING THE WEST AMERICAN TERTIARY GEOLOGICAL SCALE.

A summation of results obtained in a study of the position and relationships of the zone containing a *Merychippus* fauna in the North Coalinga region of California may be stated as follows:

1. The Merychippus zone is evidently included within the limits of the Monterey series of California as now defined.

2. The Merychippus zone evidently represents the faunal stage of the Turritella ocoyana zone, and the stratigraphic stage of the "Temblor" beds of F. M. Anderson.

3. The fauna of the Merychippus zone is not older than the stage of the Middle Miocene as represented by the Mascall and Virgin Valley faunas of the Great Basin region, and by the Pawnee Creek stage of the Great Plains area.

4. It seems inconceivable that the fauna of the Merychippus zone can represent any stage of the Oligocene of the mammal-bearing beds of western North America. The possibility that it represents any portion of the mammal-bearing Lower Miocene of this continent carries with it the supposition of wide divergence of faunas in different areas of America in Lower Miocene time.

5. The Mascall and Virgin Valley formations of the Great Basin region are of approximately the same age or older than the "Temblor."

6. The lack of adjustment between the time scale of the California area of the Pacific Coast Marginal Marine province and that of the Great Basin province suggests that correlation dependent on use of percentages of marine molluscan species as commonly defined has tended to place time divisions relatively too far from the present, in comparison with the original time scale of Europe. The lack of adjustment suggests further that the Middle Miocene mammal-bearing beds of North America should possibly occupy a lower position in the scale, in comparison with the scale of Europe, than that to which they have generally been assigned.

JACALITOS FAUNA.

Within the limits of the beds mapped as Jacalitos a few vertebrate remains representing *Pliohippus?* were reported by Arnold and Anderson.[1] Additional material of this horse and specimens representing a *Neohipparion* species have been found by investigators since that time.

[1] Ralph Arnold and Robert Anderson, U. S. Geological Survey Bull. 398, p. 300, and pl. 33, figs. 3 and 3a, 1910.

PLIOHIPPUS OR PROTOHIPPUS.

One specimen figured by Arnold and Anderson was determined by J. W. Gidley as *Pliohippus*. It represents a species of medium size with somewhat flattened protocone, and wide fossettes with slightly crinkled enamel walls. (See Figs. 28a and 28b.) Two teeth of a type similar to that from the Jacalitos figured by Arnold and Anderson were presented to the University by F. M. Anderson. They are two upper molars rather strongly curved, and with the protocone connected with protoconule. They were obtained in Sec. 35, T. 19 S., R. 15 E., M. D. B. and M. Section 35 is largely covered by the upper portion of the Jacalitos, with Etchegoin on the east side. The chances favor the occurrence of these teeth in the upper portion of the Jacalitos.

Fragments of horse teeth found by the University party in December, 1913, at three additional localities in the area mapped as Jacalitos represent advanced types of horses. All of these localities are at the base of the Jacalitos.

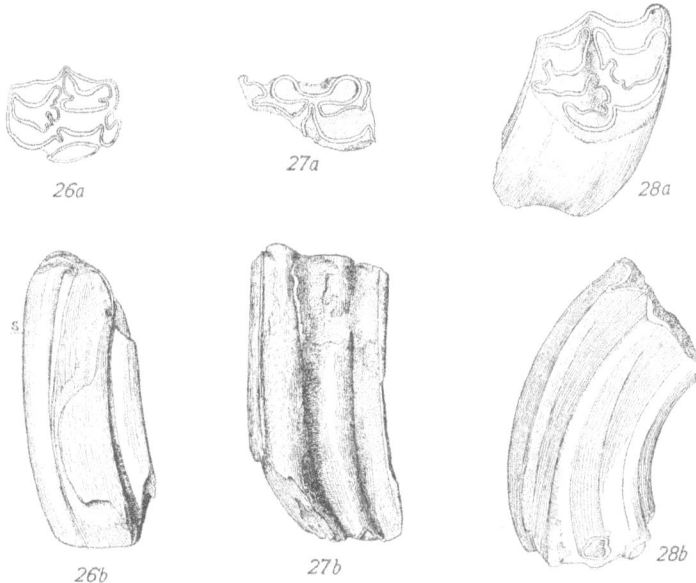

FIGS. 26a and 26b. *Neohipparion molle* Merriam M²?, no. 21370, natural size.   Lower Jacalitos, North Coalinga region California.   Fig. 26a, section of tooth at point indicated by *s* on Fig. 26b; Fig. 26b, inner view.

FIGS. 27a and 27b. *Neohipparion*, sp. M$_{\overline{2}}$?, no. 21369, natural size.   Lower Jacalitos, North Coalinga region, California. Fig. 27a, occlusal view; Fig. 27b, outer side.

FIGS. 28a and 28b. *Pliohippus*?, sp. M²., no. 165665, U. S. National Museum, natural size.   Jacalitos formation, North Coalinga region, California. Adapted from Arnold and Anderson. Fig. 28a, occlusal view; Fig. 28b, posterior ? view.

NEOHIPPARION.

An interesting upper molar three (Figs. 26a and 26b, no. 21370, locality 2076) found by J. H. Ruckman at the base of the Jacalitos, but below the line as mapped by Arnold and Anderson, represents a species of *Neohipparion*. The much flattened protocone of this tooth is widely separated from the protoconule. The fossettes are rather narrow, and their walls are only moderately plicated.

Compared with a number of practically perfect specimens of M^3 in the *Hipparion* collection from the Ricardo fauna of the Mohave area, this tooth is somewhat longer, the walls of the fossettes show less plication, the protocone is absolutely wider anteroposteriorly and much narrower transversely. The species represented by specimen 21370 is evidently distinct from the Ricardo forms, and from all other described *Hipparion* species of the Pacific Coast and Basin provinces. Whether it is a more or a less advanced species than the Mohave form is not entirely clear. The slightly greater length of crown, and the large, much-flattened protocone may indicate a more advanced stage in the Coalinga species. The *Neohipparion* species represented by no. 21370 is described as *Neohipparion molle*. This species is characterized by length and narrowness of upper molar crown, simplicity of enamel borders of the narrow fossettes, and unusually large anteroposterior diameter of the laterally compressed protocone.

A fragmentary long-crowned lower molar (no. 21369, Figs. 27a and 27b) from locality 2126 is very narrow transversely. The metaconid-metastylid column is very long anteroposteriorly, while the inner groove on this column is wide and deep. The anteroexternal region shows the sharp angle commonly seen in *Hipparion*. The fold separating the protoconid and hypoconid inserts its inner end between the folds separating protoconid and metaconid and hypoconid and metastylid. This tooth corresponds in many characters to DM_3 of *Equus occidentalis*, but is much too long to represent a temporary tooth. The characters of this tooth correspond in general to those of *Hipparion*, but it does not resemble the forms from the Ricardo fauna. In the Coalinga specimen the metaconid-metastylid column is wider anteroposteriorly, the groove between the metaconid and metastylid is wider, and the characters in general are suggestive of a more specialized form than those of Ricardo.

The Coalinga form seen in no. 21369 most nearly resembles a *Hipparion* or *Neohipparion* type quite certainly derived from the Rattlesnake Pliocene of the John Day region in eastern Oregon. Specimen 554 in the University collections from the John Day Valley was found in a basin containing both Rattlesnake Pliocene and Mascall Miocene, but the matrix covering the specimen is like the reddish-brown Rattlesnake beds, and very different from the white or gray Mascall beds of this area. The $M_{\overline{3}}$ of this Rattlesnake specimen has dimensions near those of the Coalinga specimen. The metaconid-metastylid column shows a somewhat flatter outer groove in no. 21369, but this may be due to difference in wear. The Rattlesnake specimen shows a very prominent fold on the anterior side of the hypoconid. The corresponding region of the Coalinga specimen is broken away, but there is a sudden bend in the small portion of the wall remaining at this point suggesting the presence of this fold.

A *Hipparion* or *Neohipparion* form from the Thousand Creek beds of northern Nevada shows characters near those of the Rattlesnake form, and suggests the structure in no. 21369 from the Coalinga region, but these resemblances do not necessarily mean specific identity, and the Thousand Creek form may represent a distinct type.

It is perhaps significant that both the upper and the lower teeth referred to the genus *Hipparion* or *Neohipparion* suggest a type different from that of the well-known Ricardo species, and possibly more advanced. The Jacalitos species seems most nearly related to types of the Rattlesnake beds.

MEASUREMENTS.

	No. 21370 Coalinga.	No. 21311 Ricardo. Unworn.	No. 21369 Coalinga.	No. 19847 Ricardo.	No. 554 Rattlesnake.	No. 19414 Thousand Creek.
M^3, anteroposterior diameter	19.3 mm.[1]	21				
M^3, transverse diameter	15.8[1]	17.5				
M^3, length of crown	48	38				
M^3, anteroposterior diameter of protocone at middle height of crown	9.4	7.3				
$M_{\overline{3}}$, greatest anteroposterior diameter			23.4	24.5	25.9	29
$M_{\overline{3}}$, transverse diameter			11.4	9	10.5	11
$M_{\overline{3}}$, length of crown			48 worn	48.9	57.5 worn	66
$M_{\overline{3}}$, anteroposterior diameter of metaconid-metastylid column			15	11.1	14.4	14.6

AGE OF JACALITOS VERTEBRATE FAUNA.

The Jacalitos fauna as now known is characterized by the presence of *Neohipparion* occurring only in the lowest beds, and by *Pliohippus* or *Protohippus* apparently occurring a little higher than the *Neohipparion* specimens in the basal portion of the section. Specimens resembling the *Pliohippus?* of this formation are presumed to come also from the upper portion of the section.

[1] Approximate.

The time relations of the Jacalitos vertebrates to the faunas of the Great Basin Province are not entirely clear, but the closest relationships seem to be with the Lower Pliocene.

FAUNAS OF THE ETCHEGOIN AREA.

Vertebrate remains were found at numerous localities in the area mapped as Etchegoin. The forms obtained fall naturally into two groups. One fauna was found at a series of localities near the base of the section, the localities being in or above the zone of the pelecypod *Glycimeris coalingensis*. The other localities are scattered over an area underlain by the upper portion of the Etchegoin of this section.

PLIOHIPPUS COALINGENSIS ZONE.

At several localities near the base of the Etchegoin section in this region, a fauna is found including remains of horses of a considerably more advanced type than those of the Jacalitos. The most characteristic form is *Pliohippus coalingensis*. This fauna is best known from localities 2073, 2074, 2078, and 2090. It comprises the following forms:

> *Pliohippus coalingensis* (Merriam).
> *Pliohippus?*, sp. small.
> *Procamelus?*, sp.
> *Platygonus?*, sp.

The faunal zone near the base of the Etchegoin in the North Coalinga region represents a distinctly more advanced stage than that of the Jacalitos.

PLIOHIPPUS COALINGENSIS (Merriam, J. C.).

Protohippus coalingensis Merriam, J. C., *Science*, N. S., vol. 40, p. 645, August, 1914.

Type specimen no. 21341 from locality 2073. Lower Etchegoin ten miles north of Coalinga, California. A species of medium size, approaching the dimension of *Pliohippus supremus*. The type specimen (Figs. 29a and 29b) differs from the type specimen of *P. supremus* in the heavier mesostyle, relative narrowness of the crown even toward the base, narrower fossettes, much less complicated enamel walls of the fossettes, and much smaller more nearly circular protocone.

The nearest approach to this form among the Equidae known west of the Wasatch is *Pliohippus fairbanksi*, a species imperfectly represented in the Ricardo beds of the Mohave Desert. The lower Etchegoin form resembles the Ricardo species approximately in size, in simplicity of the fossettes, in the very small, round protocone, and weak connection of the protoloph and metaloph. The crown of the Etchegoin form is, however, less curved and is much narrower.   The narrowness of the Coalinga form is apparently a normal character of worn teeth.   It may be that later collections will show that these two forms are specifically identical, or at most not removed

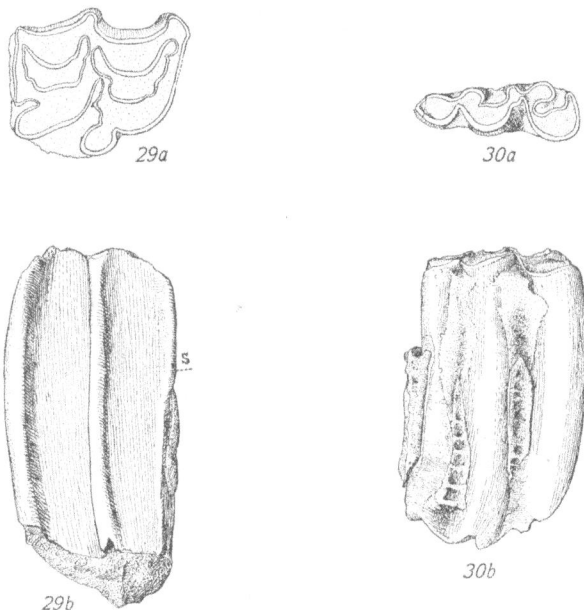

FIGS. 29a and 29b. *Pliohippus coalingensis* (Merriam). P$\underline{4}$, type specimen, no. 21341, natural size. Etchegoin formation, North Coalinga region, California. Fig. 29a, cross-section of tooth at point indicated by *S* on Fig. 29b; Fig. 29b, outer view.

FIGS. 30a and 30b. *Pliohippus?*, sp. M$\overline{3}$, no. 21362, natural size. Etchegoin formation, North Coalinga region, California. Fig. 30a, occlusal view; Fig. 30b, outer view.

farther than by subspecific characters.   The imperfectly known *Pliohippus* species of the Thousand Creek beds seems to have a somewhat different form and differently constructed fossettes.

Pliohippus coalingensis is somewhat larger and more advanced than the *Pliohippus* or *Protohippus* species of the Jacalitos.

MEASUREMENTS.

	No. 21341 Coalinga.	No. 19789 Ricardo.	P. supremus Type.
P$\underline{4}$, transverse diameter	24.4 mm.	26	
P$\underline{4}$, anteroposterior diameter	27.2	25	
P$\underline{4}$, height of crown	53	55	
M$\underline{1}$, transverse diameter			25
M$\underline{1}$, anteroposterior diameter			24

PROTOHIPPUS OR PLIOHIPPUS, sp.

Associated with *P. coalingensis* at locality 2074 are fragmentary remains of teeth evidently representing a species near the *Pliohippus?* of the Jacalitos.   The material available is not sufficient to permit definite determination of the affinities of the form. At locality 2090 a lower molar, no. 21362 (Figs. 30*a* and 30*b*), represents a species that may be identical with the smaller protohippine form at locality 2074.

CAMELIDÆ.

A single fragment of a metapodial (Fig. 31) from locality 2090 represents a camel approximating in size the large species of *Procamelus*.

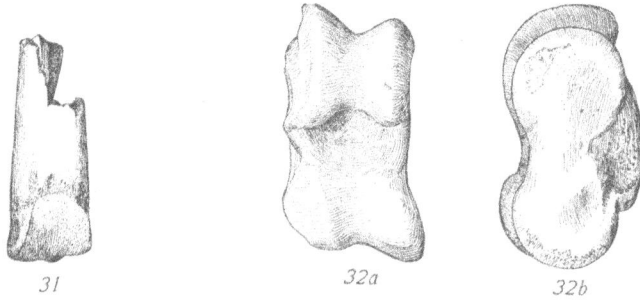

FIG. 31. Camelid.   Portion of distal end of metapodial.   No. 21363, × ½.   Etchegoin formation, North Coalinga region, California.

FIGS. 32*a* and 32*b*. Astragalus of *Platygonus*-like form.   No. 21371, natural size.   Etchegoin? formation, North Coalinga region, California.   Fig. 32*a*, superior view; Fig. 32*b*, medial view.

SUIDÆ.

A single astragalus (no. 21371, Figs 32*a* and 32*b*) from locality 2123 in the lower Etchegoin zone represents a peccary-like form.   In dimensions and form it is near *Platygonus*.   It is possible that this specimen belongs with the latest fauna from this region.

LATER FAUNA FOUND ON UPPER PORTION OF ETCHEGOIN AREA.

Vertebrate remains were found in considerable quantity on exposures representing the upper portion of the Etchegoin section in the North Coalinga region. The members of the University party all obtained specimens in this area, but the largest part of the collection was secured by J. O. Nomland, who was engaged in making a special study of the Etchegoin and its relation to adjacent formations. The collections from this area were obtained from a considerable number of localities situated mainly in a zone ranging from one-half mile to a little more than a mile east of the Glycimeris zone at the base of the Etchegoin.   The only exception to the rule is the possible representation of the fauna by a peccary astragalus at locality 2123 near the Glycimeris zone.   The localities range in elevation from near 600 feet to a little more than 900 feet above sea level.   At a number of localities this fauna is associated with invertebrates of several horizons representing the upper portion of this Etchegoin section.   The vertebrate specimens were not found in place, but appeared on exposures which seemed to consist solely of Etchegoin material.

As the later fauna in general resembles that of the Pleistocene as closely as it does any known Pliocene of America, the writer made special inquiry as to the possibility of derivation of the specimens from Pleistocene accumulations.   Terrace deposits bordering the Great Valley are reported in the region to the north of Coalinga at several levels, some of which are considerably above that at which the collections representing the later fauna were made.   The bones were all obtained on hills seemingly consisting entirely of Etchegoin, and with no suggestion of terrace deposits.   However, it is known that important terraces in the region to the north may have an exceeding small amount of material left upon them, so that the rarity of loose material may not indicate absence of terraces.   Members of the University party who collected vertebrate remains in this area state that there are terraces at levels both above and below the localities at which bones were collected, and that the hills upon which the vertebrate material was obtained seem to have been formed by erosion since the upper terraces were completed, while the lower terraces do not reach up to the level of these hills.

It does not appear to the writer that final evidence as to the occurrence of this fauna is now before us.   Since the question is of importance it is desirable to continue the examination of this area with a view to obtaining material in place.

The fauna found in the upper Etchegoin area includes the following forms:

> *Equus* or *Pliohippus*, sp.
> *Camelops* or *Pliauchenia*, sp.
> *Procamelus?*, sp.
> *Cervus* or *Odocoileus*, sp.
> *Tayassu* or *Mylohyus?*, sp.
> Mastodon.
> *Testudo?*, sp.
> Fish vertebræ, and scattered bulbous fish bones.

EQUUS or PLIOHIPPUS, sp.

The equid remains known from the upper Etchegoin area include upper and lower molars, the astragalus, and all of the phalanges.   The molars are very large, exceeding slightly the dimensions of the largest specimens of *Pliohippus* available for comparison.   The form of the upper cheek-teeth (Figs. 33a to 34b) strongly resembles

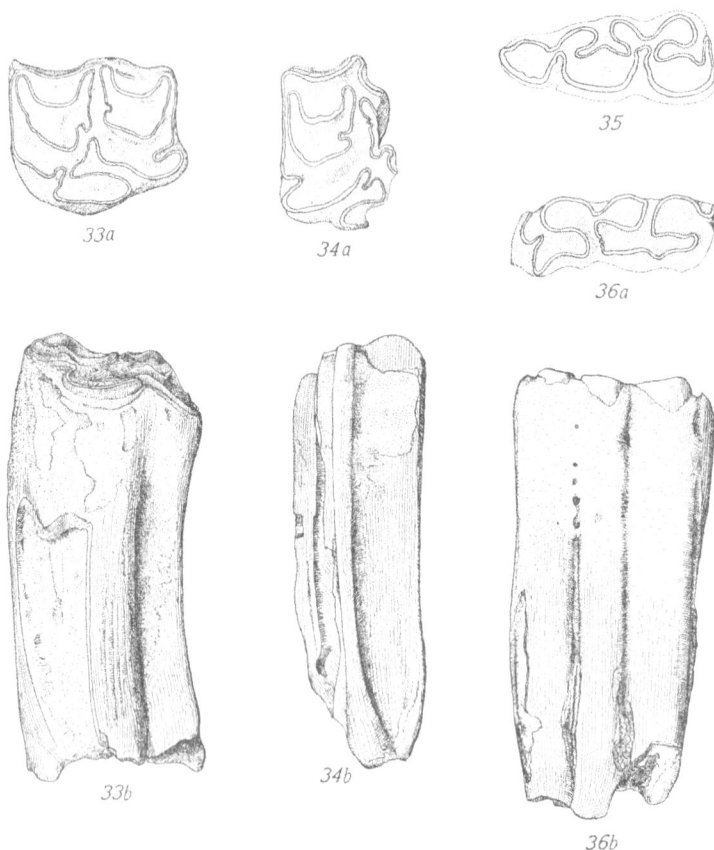

FIGS. 33a and 33b.   *Equus* or *Pliohippus*, sp. M^1, no. 21330, natural size.   From upper Etchegoin area, North Coalinga region, California.   Fig. 33a, occlusal view, outer enamel wall absent; Fig. 33b, inner view.

FIGS. 34a and 34b.   *Equus* or *Pliohippus*, sp. P^4?, no. 21331, natural size.   From upper Etchegoin area, North Coalinga region, California.   Fig. 34a, occlusal view; Fig. 34b, outer view.   ·

FIG. 35.   *Equus* or *Pliohippus*, sp. M$_{\overline{3}}$, occlusal view.   No. 21333, natural size.   From upper Etchegoin area, North Coalinga region, California.

FIGS. 36a and 36b.   *Equus* or *Pliohippus*, sp. M$_{\overline{1}}$?, no. 21332, natural size.   From upper Etchegoin area, North Coalinga region, California.   Fig. 36a, occlusal view; Fig. 36b, inner view.

that of *Equus*.   The protocone differs from that of all *Pliohippus* species known to the writer in its relatively great anteroposterior diameter.   Suggestions of *Pliohippus* are seen in the simplicity and unusual width of the fossettes, in the somewhat more slender mesostyle, and in the convex inner face of the protocone. The protocone is also short anterior to its union with the protoconule. The general form of the protocone may, however, be approximated in *Equus*.

MEASUREMENTS.

mm.

P^4?, transverse diameter (no. 21331) 29
M^1, anteroposterior diameter (no. 21330).. 30
M^1, anteroposterior diameter of protocone (no. 21330)................................ 8.8

The lower cheek-teeth (Figs. 35 to 36b) are large, long, and heavily cemented. The dimensions approximate those of the largest specimens of *Equus occidentalis* from the Pleistocene of California. The crowns are larger and relatively narrower than in any *Pliohippus* available to the writer for comparison. The metaconid-metastylid column is divided internally by a deep, wide groove, which approaches more nearly the normal form in *Equus* than to that in typical *Pliohippus*.

MEASUREMENTS.

mm.

No. 21332.
Lower cheek-tooth, anteroposterior diameter................................. 34.1
No. 21333.
M$_{\overline{1}}$, anteroposterior diameter... 27.5
M$_{\overline{3}}$, anteroposterior diameter... 36.2
M$_{\overline{3}}$, transverse diameter... 13.8

The limb elements (Fig. 37), including the astragalus, metapodials, and the phalanges, are somewhat smaller than in average specimens of *Equus caballus*, or in the Pleistocene *Equus occidentalis*. The form of the limb elements is much as in *E. caballus*. If it differs appreciably from modern species, the difference is in the more slender form of the Coalinga specimens. The proximal end of a metacarpal shows the facets as in *Equus*. The distal keels of several metapodials are large and prominent. In one specimen representing an ungual phalanx the lateral wing of the hoof is somewhat smaller than in *Equus caballus*.

A summation of the characters of the equid form from the upper Etchegoin area shows that it very closely approaches the *Equus* group. If, as is suggested by certain characters of the upper molars, there is close affinity with *Pliohippus*, this would seem the most advanced species thus far known in that group, and one presumably furnishing a transition to *Equus*.

PROBOSCIDEAN REMAINS.

A considerable number of fragments representing a mastodon-like form have been found at several localities in the upper Etchegoin area. The greater part of a tooth (no. 21339, Fig. 38) was found at locality 2119, and another tooth of different

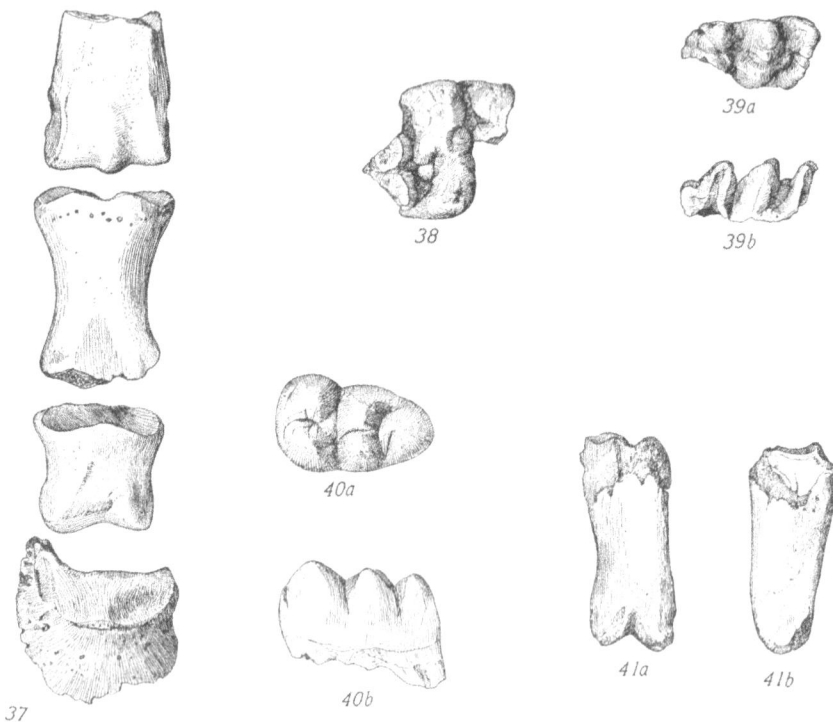

Fig. 37.   *Equus* or *Pliohippus*, sp.   Anterior view of distal end of third metapodial, and phalanges of third digit.   No. 21380 (not all elements known to represent one individual) × ⅓.   From upper Etchegoin area, North Coalinga region, California.

Fig. 38.   Portion of a mastodon cheek-tooth seen from above.   No. 21339 × ¼.   From upper Etchegoin area, North Coalinga region, California.

Figs. 39a and 39b.   Portion of a mastodon cheek-tooth.   No. 21340 × ½.   From the upper Etchegoin area, North Coalinga region, California.   Fig. 39a, occlusal view; Fig. 39b, lateral view.

Figs. 40a and 40b. $M_{\overline{1}}$ of *Mylohyus*-like form.   No. 21360, natural size.   From upper Etchegoin area, North Coalinga region, California.   Fig. 40a, occlusal view; Fig. 40b, side view.

Figs. 41a and 41b.   Proximal phalanx of *Platygonus* or *Mylohyus*-like form.   No. 21338, natural size.   From upper Etchegoin area, North Coalinga region, California.   Fig. 41a, superior view; Fig. 41b, side view.

form (no. 21340, Figs. 39a and 39b) at locality 2079.   The first specimen represents a very large mastodon, which is, so far as determinable, not far from the stage of development of the Pleistocene species.   The second specimen apparently represents a smaller form with narrower teeth.   Neither form is necessarily excluded from the types of mastodon occurring in the Pliocene.

SUIDÆ.

Two large third lower molars represent a peccary-like type not previously known in the Pacific Coast region.   The tubercles are low, blunt-conical as in *Tayassu* and *Mylohyus*, and are not distinctly connected by transverse crests as in *Platygonus*. Intermediate tubercles do not appear as well developed as in the *Tayassu* and *Mylohyus* specimens examined by the writer.   In size the larger specimen (Figs. 40a and 40b) corresponds approximately to *Mylohyus*, sp. a from Conard Fissure, as figured by Brown.[1]

This form is near *Tayassu* or *Mylohyus*.   If other specimens show somewhat advanced development of the secondary tubercles, it may represent *Mylohyus*.   The species is probably new.

MEASUREMENTS.

	No. 21360, mm.	No. 21359.
$M_{\overline{3}}$, greatest anteroposterior diameter	27.2	24.3
$M_{\overline{3}}$, transverse diameter	15.9	13.4

A single proximal phalanx (no. 21338, Figs. 41a and 41b) from locality 2079 represents a form near *Platygonus*, but the element is a little less enlarged inferiorly than that genus.   This difference is possibly due in part to wear on the fossil specimen.

CAMELIDÆ.

Remains representing two groups of camels are found in the upper Etchegoin area.   One contains forms near *Camelops* or *Pliauchenia*.   The other type is not more than half as large as the first, and is possibly a representative of a large species of *Procamelus*.   The two species were found at locality 2119.

The first form is represented by large metapodial fragments (nos. 21334 and 21379, Figs. 42a to 43).   A proximal phalanx and several fragmentary bones possibly belong also to this type.   This animal was near the size of *Camelops hesternus* of the California Pleistocene.   The metapodial fragments (no. 21334 and 21379) of the large camel represent the proximal end of an anterior cannon-bone, and the distal portion of an element which is possibly from the hind limb.   The proximal end of the anterior cannon-bone is closely similar to that of *Camelops*.   The main difference between this specimen and that of *Camelops hesternus* from Rancho La Brea suggested is in the slightly fuller bridge of bone uniting metacarpals three and four posteriorly

[1] Brown, B., *Mem. Amer. Mus. Nat. Hist.*, vol. 9, pl. 24, 1908.

at the proximal end.   If the fragment of the distal end of a cannon-bone represents an anterior limb, this element is more slender than in *Camelops hesternus.*

A proximal phalanx (Figs. 45a and 45b, no. 21356) from locality 2370 is longer and much more slender than the phalanges of the *Camelops* species from Rancho La

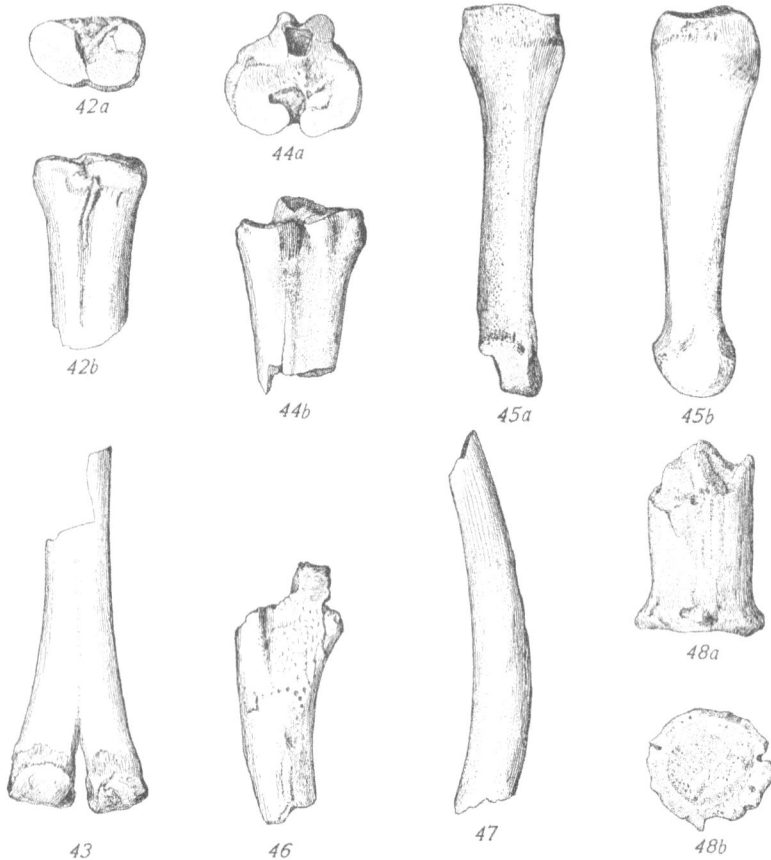

FIGS. 42a–43.   Metapodials of large *Camelops*-like form.   From upper Etchegoin area, North Coalinga region, California. Figs. 42a and 42b.   Proximal end of anterior cannon-bone.   No. 21334 × ¼.   Fig. 42a, proximal view; Fig. 42b, anterior view. Fig. 43.   Anterior view of distal end of metapodial, possibly of posterior limb.   No. 21379 × ¼.

FIGS. 44a and 44b.   *Procamelus*?, sp.   Proximal end of cannon-bone from posterior limb.   No. 21335 × ⅓.   From upper Etchegoin area, North Coalinga region, California.   Fig. 44a.   Proximal view; Fig. 44b, anterior view.

FIGS. 45a and 45b.   *Pliauchenia*?.   Proximal phalanx.   No. 21356 × ⅓.   From upper Etchegoin area, North Coalinga region, California.   Fig. 45a, superior view; Fig. 45b, side view.

FIGS. 46–48b.   *Cervus* or *Odocoileus*, sp.   From upper Etchegoin area, North Coalinga region, California.   Fig. 46. Portion of flattened forked antler.   No. 21336 × ⅓.   Fig. 47.   Portion of slender tine with round cross-section.   No. 21337 × ⅓.   Figs. 48a and 48b.   Base of antler with burr.   No. 21357 × ⅓.   Fig. 48a, lateral view; Fig. 48b, basal view.

Brea, or than in the large camel of the Pleistocene Manix beds in the Mohave area. It evidently represents a large species quite different from the Pleistocene *Camelops* forms, and from all other camels known in the Pacific Coast Pleistocene. It possibly corresponds to *Pliauchenia* or to some genus other than *Camelops*.

The second camelid form known from the upper Etchegoin area is represented by the proximal end of a posterior cannon-bone (no. 21335) of an adult individual somewhat larger than the Recent llama. This animal was less than half the bulk of *Camelops hesternus* or of the *Camelops*-like form represented by specimen no. 21334. The form of the proximal end of the cannon-bone (Figs. 44*a* and 44*b*) is entirely different from that of *Camelops* and *Camelus*. It is much narrower transversely, and the posterior facet for the cuboid is much larger than in *Camelops*. The form is near that of *Procamelus*, though the transverse diameter is a little larger than in any *Procamelus* species available for comparison. The facets for the cuboid and mesocuneiform are lunate instead of oval as in *Procamelus*.

CERVIDÆ.

At several localities (no. 2119, 2079, 2370 and 2374) in the upper Etchegoin area fragments of antlers have been found representing forms of a cervid type. Two specimens consist of the basal portion with the burr of large antlers that had been shed. They evidently represent a very advanced form of the true cervid type. The larger specimen (Figs. 48*a* and 48*b*, no. 21357) is from an individual with antlers approximating in size those of large specimens of the Recent *Cervus* or *Odocoileus*.

One specimen (no. 21336, Fig. 46) consists of a flattened beam apparently dividing above into two branches of different dimensions. The shape of the beam is like that in *Cervus*, and resembles a form that may occur in *Merycodus*. It is larger than in specimens of the nearest forms of *Merycodus* known to the writer, and the branches above are more uneven in size than in *M. necatus*, the nearest merycodont. Another specimen (no. 21337, Fig. 47) is a portion of a long, slender, gently curved tine or branch with nearly circular cross-section. The convex side of the tine shows a double row of very small, irregular tubercles separated by a faint groove. This tine is larger than that of any merycodont examined by the writer.

The evidence before us indicates that the specimens from this fauna represent a true cervid of the *Odocoileus* or *Cervus* type.

TESTUDINATE REMAINS.

A fragment of a carapace representing a very large tortoise-like form (no. 21378, Figs. 49*a* and 49*b*) was obtained with the collection found in the upper Etchegoin area. The species is unlike any form known in this region at the present time.

AGE' OF LATER FAUNA FROM UPPER ETCHEGOIN AREA.

The aspect of the later fauna from the Etchegoin area resembles in some features that of the Pleistocene. The horse in this assemblage seems more advanced than any American Pliocene form known to the writer. *Equus simplicidens* Cope from the Blanco Pliocene of Texas is considered by Gidley[1] to represent *Pliohippus*. The protocone of the Coalinga species is very large, long anteroposteriorly, compressed laterally, and nearly flat on the inner side. In Cope's figure of *E. simplicidens* the protocone is incomplete, but seems smaller than in the Coalinga form, and is restored as smaller by Gidley.

The metapodial of the larger camel in this fauna approaches that of a Pleistocene *Camelops*, but the large phalanges, apparently representing this fauna, do not correspond to any of the Pleistocene camel species known in the Pacific Coast region. The smaller camel seems more nearly to represent a Pliocene *Procamelus* or some allied

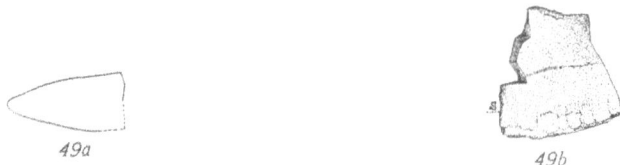

FIGS. 49a and 49b. Portion of carapace of a large tortoise. No. 21378 × ½. From the upper Etchegoin area, North Coalinga region, California. Fig. 49b, superior view of marginal? plate, lateral margin to left of figure; Fig. 49a, transverse section at s, at lower end of Fig. 49b.

group. The cervid remains represent deer of a distinctly modern type. Nothing of this character has been known in the Pliocene of America so far as the writer is aware. The peccary is of an advanced type and is near the Pleistocene stage of development. The mastodon remains are fragmentary, and little weight has been placed on them in distinction between Pliocene and Pleistocene. This is true also of the tortoise, and of the fish remains.

In attempting to fix the age of the fauna it is first necessary to determine whether the collection available comes from one source or whether it is derived from two or more sets or deposits. It is possible that one part comes from Pleistocene terraces, and that other elements, such as the *Procamelus*, are from the Etchegoin formation. The fact that both the small camel, and the large species show difference from the known Pleistocene forms naturally suggests that a really new faunal assemblage is present here. However, it will be difficult to give satisfactory evidence that this is a faunal unit until some of the forms with a distinctly Pleistocene aspect have

[1] Gidley, J. W., *Bull. Amer. Mus. Nat. Hist.*, vol. 14, p. 124, 1901.

been found in place in the Etchegoin, or until some of those representing an older type, as the *Procamelus*-like form, have been discovered in terrace deposits.

If the fauna is a unit and occurs in place in the Etchegoin, it indicates either that the Etchegoin is younger than the Miocene or Pliocene stage which it has been presumed to represent, or that a modern type of fauna has appeared relatively early in California.

The Etchegoin, upon the outcrops of which the odd Cervus fauna is found, is a formation of considerable thickness. It is covered by the Tulare, which is generally considered to be represented by much more than 2,000 feet of strata. Since the close of Tulare time important events referred to the Pleistocene have taken place in this region. From purely physical studies there seems good reason for believing that the upper portion of the Etchegoin represents an epoch separated from Recent time by a considerable period.

The invertebrate fauna of the upper portion of the Etchegoin area from which the vertebrate remains were obtained consists, according to Mr. Nomland, of about forty-nine species of which twenty are extinct. Arnold and Anderson[1] considered the Etchegoin as of Miocene age. The total fauna was noted to contain 65 per cent. of extinct species. Arnold and Anderson were in agreement with F. M. Anderson that the Etchegoin is to be correlated with the San Pablo formation. Arnold and Anderson also considered the upper Etchegoin as the equivalent of the lower Purisima. The most recent studies of Bruce Clark, Bruce Martin, and J. O. Nomland, show that the Etchegoin fauna is distinctly more recent than that of the San Pablo. The suggestion that the Etchegoin was contemporaneous with the San Pablo, but of a different geographic phase, seems not to be adequate in view of the fact that the San Pablo fauna closely matches that of the "Santa Margarita" occurring in the second formation below the Etchegoin, and separated from it by the stage referred to the Jacalitos in the North Coalinga region.

From the lower portion of the Tulare formation above the Etchegoin, Arnold and Anderson[2] list thirteen species, of which all but three are considered as extinct forms. It was suggested by these authors that the basal portion of the Tulare represented the lower Pliocene, but that the upper portion extended into the Pleistocene.

It appears to the writer improbable that the Etchegoin can be included in the Miocene. The assemblage of evidence from a study of the invertebrate faunas, interpreted in terms of the Lyellian percentage method as it was understood when that method was proposed, suggests Pliocene age. This was the determination given by F. M. Anderson in the original description of the Etchegoin.

[1] Arnold, R., and Anderson, R., U. S. Geol. Surv. Bull. 398, p. 139, 1910.

[2] Arnold, R., and Anderson, R., U. S. Geol. Surv. Bull. 398, p. 154, 1910.

If the vertebrates obtained from the upper portion of the Etchegoin area in the North Coalinga region were actually derived from Etchegoin strata, we apparently have a relatively advanced fauna in beds which according to evidence from geology and invertebrate palæontology suggest Pliocene age. This relation of the evidence from vertebrate and invertebrate palæontology is in a manner similar to that in the lower portion of the Coalinga geologic section, where the "Temblor" invertebrates are determined as older than the vertebrate Merychippus fauna. The fact that the relation is the same in the Etchegoin area as in the "Temblor" area may suggest that the vertebrate fauna actually occurs in the Etchegoin strata. If this is true, the balance of evidence would seem to the writer to indicate that the strata in question are not older than middle Pliocene.

Alternative possibilities of explanation that remain open are the suggestion that this peculiar fauna is derived from vestiges of terraces that reached only across the outer or upper portion of the Etchegoin area, or that it is composed in part of Pleistocene terrace material and in part of Pliocene forms derived from strata of the Etchegoin formation. Although the field work already done has been carefully conducted, the final settlement of this important problem seems to require still more intensive study.

TABLE REPRESENTING POSITION OF FORMATIONS CONTAINING VERTEBRATE FAUNAS IN NORTH COALINGA REGION.

The following table gives approximate position of beds containing land vertebrate faunas in the North Coalinga region in comparison with the situation of the principal divisions of the geological scale in the Great Basin and California regions.

Geological Periods.		Local Formations, California Area.	Local Formations, Great Basin Province.	Faunal Zones.	
				Vertebrate Faunas.	Invertebrate Faunas.
Pleistocene		Rancho La Brea	Manix	Smilodon ⎫ Equus ⎬ Stages not Camelops ⎬ defined Elephas ⎭	Echinarachnius excentricus Turritella jewetti
Pliocene		Tulare Etchegoin Jacalitos	 Thousand Creek Rattlesnake Ricardo	Hyaenognathus Ilingoceras Pliohippus Hipparion	Echinarachnius gibbsii
Miocene	Upper	San Pablo ("Santa Margarita")	Barstow Beds	{ Merychippus calamarius Merycodus necatus	Astrodapsis
	Middle	Monterey ("Temblor")	Mascall and Virgin Valley	{ Merychippus isonesus Parahippus Dromomeryx	Turritella ocoyana
	Lower	?"Vaqueros"	Columbia Lava	No known fauna	Turritella inezana
Oligocene	Upper Middle Lower	San Lorenzo	Upper John Day Middle John Day ?Lower John Day	Promerycochoerus Eporeodon Imperfectly known fauna	Turritella diversilineata
Eocene	Upper	Tejon	Upper Clarno	No fauna (Flora—Sequoia heeri, Betula, Alnus, Quercus, Acer, Ficus)	Turritella merriami Turritella uvasana
	Lower	Martinez	Lower Clarno	No fauna (Flora—Lygodium, Asplenium, Equisetum, Juglans, Magnolia)	Turritella pachecoensis Turritella infragranulata

LIST OF IMPORTANT LOCALITIES AT WHICH VERTEBRATE SPECIMENS WERE COLLECTED IN THE NORTH COALINGA REGION.

All Townships and Ranges referred to Mount Diablo Base and Meridian.

MERYCHIPPUS ZONE—
> Locality 2124: SW. ¼ sec. 28, T. 18 S., R. 15 E.

JACALITOS—
> Locality 2126: NE. ¼ of NE. ¼ of sec. 15, T. 19 S., R. 15 E.
> Locality 2076: SW. ¼ of sec. 34, T. 18 S., R. 15 E.
> Locality 2077: NE. ¼ sec. 15, T. 19 S., R. 15 E.

ETCHEGOIN, PLIOHIPPUS COALINGENSIS ZONE—
> Locality 2073: SW. ¼ sec. 13, T. 19 S., R. 15 E.
> Locality 2074: SW. ¼ sec. 12, T. 19 S., R. 15 E.

Locality 2078: NW. ¼ of SW. ¼ sec. 13, T. 19 S., R. 15 E.
Locality 2090: NW. ¼ sec. 13, T. 19 S., R. 15 E.

FAUNA FROM UPPER ETCHEGOIN AREA:
Locality 2079: the NW. ¼ of NW. ¼ sec. 19, T. 19 S., R. 16 E.
Locality 2080: NW. ¼ of SW. ¼ sec. 6, T. 19 S., R. 16 E.
Locality 2082: SW. ¼ of NE. ¼ sec. 1, T. 19 S., R. 15 E.
Locality 2084: north line of SE. ¼ of sec. 12, T. 19 S., R. 15 E.
Locality 2086: NW. ¼ of SW. ¼ of sec. 6, T. 19 S., R. 16 E.
Locality 2089: NW. ¼ of sec. 19, T. 19 S., R. 16 E.
Locality 2119: SE. ¼ of sec. 12, T. 19 S., R. 15 E.
Locality 2368: NW. ¼ of the SW. ¼ of sec. 19, T. 19 S., R. 15 E.
Locality 2370: NE. ¼ of sec. 25, T. 19 S., R. 15 E.
Locality 2372: NE. ¼ of the SE. ¼ of sec. 25, T. 19 S., R. 15 E.
Locality 2374: W. ½ of the NW. ¼ of sec. 25, T. 19 S., R. 15 E.

www.ingramcontent.com/pod-product-compliance
Lightning Source LLC
Chambersburg PA
CBHW081333190326
41458CB00018B/5978